Developing Numeracy
SOLVING PROBLEMS
ACTIVITIES FOR THE DAILY MATHS LESSON

D1392868

year
5

Hilary Koll and Steve Mills

A & C BLACK

Contents

Problems involving measures

Answers

Reprinted 2001, 2002 (twice), 2003, 2004, 2005, 2006, 2007
Published 2000 by A & C Black Publishers Limited
38 Soho Square, London W1D 3HB
www.acblack.com

ISBN 978-0-7136-5448-6

Copyright text © Hilary Koll and Steve Mills, 2000
Copyright illustrations © Michael Evans, 2000
Copyright cover illustration © Charlotte Hard, 2000
Editors: Lynne Williamson and Marie Lister

The authors and publishers would like to thank the following teachers for their advice in producing this series of books:
Stuart Anslow; Jane Beynon; Cathy Davey; Ann Flint; Shirley Gooch; Barbara Locke; Madeleine Madden; Helen Mason;
Fern Oliver; Jo Turpin.

A CIP catalogue record for this book is available from the British Library.

Printed and bound in Great Britain by Cromwell Press Ltd, Trowbridge.

A & C Black uses paper produced with elemental chlorine-free pulp, harvested from managed, sustainable forests.

Introduction

Developing Numeracy: Solving Problems is a series of seven photocopiable activity books designed to be used during the daily maths lesson. It focuses on the third strand of the National Numeracy Strategy *Framework for teaching mathematics*. The activities are intended to be used in the time allocated to pupil activities; they aim to reinforce the knowledge, understanding and skills taught during the main part of the lesson and to provide practice and consolidation of the objectives contained in the framework document.

Year 5 supports the teaching of mathematics by providing a series of activities which develop essential skills in solving mathematical problems. On the whole the activities are designed for children to work on independently, although this is not always possible and occasionally some children may need support.

Year 5 encourages children to:

- choose and use appropriate number operations to solve problems and to use appropriate ways of calculating;
- solve mathematical problems and puzzles and to explore relationships and patterns;
- investigate a general statement about familiar numbers or shapes;
- explain a generalised relationship (formula) in words;
- explain their methods and reasoning;
- solve one-step and multi-step worded problems in areas of 'real life', money and measures;
- make simple conversions of pounds to foreign currency and to find simple percentages.

Extension

Many of the activity sheets end with a challenge (**Now try this!**) which reinforces and extends the children's learning, and provides the teacher with the opportunity for assessment. On occasions you may wish to read out the instructions and explain the activity before the children begin working on it. The children may need to record their answers on a separate piece of paper.

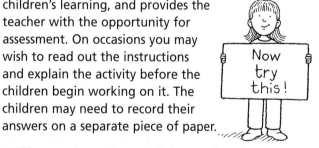

Differentiated activities

For some activities, two differentiated versions are provided which have the same title and are presented on facing pages in the book. On the left is the less challenging activity, indicated by a rocket icon: . The more challenging version is found on the right, indicated by a shooting star: . These activity sheets could be given to different groups within the class, or all the children could complete the first sheet and children requiring further extension could then be given the second sheet.

Organisation

Very little equipment is needed, but it will be useful to have available: coloured pencils, interlocking cubes, scissors, digit cards, squared paper, shape templates, protractors, calendars and small clocks.

Where calculators should be used, this is indicated on the page; otherwise it is left to the teacher's discretion.

To help teachers to select appropriate learning experiences for the children, the activities are grouped into sections within each book. However, the activities are not expected to be used in that order unless otherwise stated. The sheets are intended to support, rather than direct, the teacher's planning.

Some activities can be made easier or more challenging by masking and substituting some of the numbers. You may wish to re-use some pages by copying them onto card and laminating them, or by enlarging them onto A3 paper.

Teachers' notes

Brief notes are provided at the foot of each page giving ideas and suggestions for maximising the effectiveness of the activity sheets. These can be masked before copying.

Structure of the daily maths lesson

The recommended structure of the daily maths lesson for Key Stage 2 is as follows:

Start to lesson, oral work, mental calculation	5–10 minutes
Main teaching and pupil activities (the activities in the **Developing Numeracy** books are designed to be carried out in the time allocated to pupil activities)	about 40 minutes
Plenary (whole-class review and consolidation)	about 10 minutes

Whole-class activities

The following activities provide some practical ideas which can be used to introduce or reinforce the main teaching part of the lesson.

Making decisions

Spiders

Draw a large spider on the board with a number on it, for example *240*. Ask individual children to come to the front and write questions that give an answer of 240, for example *40 x 6*. Encourage them to describe this in a real situation, such as: *A man walked into* 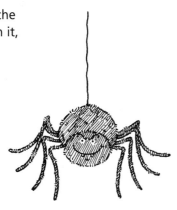 a shop and bought six magazines that each cost 40p. He paid £2.40 in total. Whole numbers, fractions or decimals could be written on the spider. The children could use any signs, including brackets, and more than two numbers in the question, for example *(50 x 4) + 40 = 240*.

Number question strip

On a strip of card or thick paper write a number fact, for example *366 + 139 = 505*. Wrap a narrow piece of paper around the strip of card so that it can slide sideways to mask one of the numbers or operator signs. Hold up the strip and ask the children to find the hidden number or sign. This can then be revealed to check that it is correct. You could build up a collection of number strips to use throughout the year.

Reasoning about numbers

Counting stick

You will need a stick which is divided into ten equal coloured sections (such as a metre stick with each 10 cm coloured). Hold the stick so that all the children can see it and point to each section along it in turn. Decide on a number (for example, eight) and ask the children to counts in eights as you point to each section. This provides practice in counting forwards and backwards and helps the children to remember the multiples of the given number.

Reasoning about shapes

Symmetrical cutting

Hold up a sheet of A4 paper and tell the children that you are going to fold the paper and make a straight cut through it. Explain that you will discard the piece you cut off. Ask the children: *Which shape might the remaining piece of paper be? What different shapes could I make?* Allow time for the children to discuss and try to visualise this with a partner. Point out that by cutting off parts of the sheet in different ways, for example, a corner, the following shapes can be made: *pentagon, concave pentagon, hexagon, triangle*, etc. As an extension, a second fold can be introduced.

Problems involving 'real life'

Paper clips

Hold up a box of paper clips (or a similar container) and state the number of paper clips in the box. Ask a range of questions about the box, for example:
How many paper clips would there be in two boxes? Ten boxes? 100 boxes?
How many paper clips would there be if I had a full box and used 27 of them?
I use half the paper clips in the box. How many paper clips are there now?
I have a full box and give two paper clips to each child in our class. How many have I now?

Problems involving money

Shopping trip

Draw five items on the board with price labels, for example, £1.75, 48p, £1.99, £2.20, 83p. Ask the children to work out the total cost for two or more items. Choose a child to ask a question in the following way: *I went shopping and bought two (or three) things. I paid £X. What did I buy?* The rest of the class should try to work out which items were bought.

Problems involving measures

Exploring letters

Write these capital letters on the board: *H, I, E, C, F, L, T.* Ask the children, working in pairs, to choose a letter and draw it on squared paper. All letters must be made from squares and must fit within a 4 x 5 border. Each pair then describes their letter (without saying what it is) by writing the perimeter and area on the board, for example: *perimeter = 26 cm, area = 12 cm²*. The other pairs try to guess the letter.

Frantic antics

The ants have bitten the signs from the number statements!

• **Write the missing signs.**

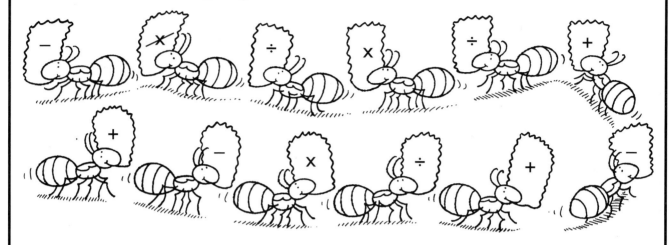

1. $18 \times 7 = 126$

2. $304 \quad 38 = 8$

3. $596 \quad 315 = 281$

4. $721 \quad 144 = 865$

5. $163 \quad 5 = 32 \cdot 6$

6. $675 \quad 25 = 27$

7. $24 \quad 9 = 216$

8. $179 \quad 165 = 344$

9. $977 \quad 8 = 985$

10. $641 \quad 178 = 463$

11. $28 \quad 8 = 224$

12. $556 \quad 229 = 327$

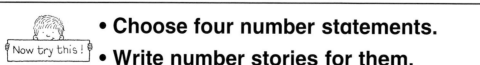

• **Choose four number statements.**
• **Write number stories for them.**

Teachers' note Encourage the children to describe how they worked out the missing signs. They could use calculators to check. Remind them that when adding or multiplying whole numbers, the largest number is the answer; and when subtracting or dividing whole numbers, the largest number is the first number. The activity 'Number question strip' on page 5 provides more practice.

**Developing Numeracy
Solving Problems Year 5
© A & C Black**

Track it down

Developing Numeracy
Solving Problems Year 5
© A & C Black

• **Write the missing digits.**

Use a calculator to help you.

1. 9   8   ×   6   4   = 6272

2. 8   ×   3   = 2400

3.   1   ×   4   = 2867

4.   ×   4   = 1596

5.   ×   3   = 598

6.   ×   9   = 2256

7.   ×   2   = 3402

8.   3   ×   = 1495

Now try this!

• **Find as many ways as you can to complete this.**

☐ + ☐ + ☐ = 1

Example: 0·1 + 0·1 + 0·8 = 1

Teachers' note Discuss how to solve calculator puzzles by looking for clues in the numbers given, for example, deciding whether it will be a large digit or a small digit, and then inserting digits and testing them. Encourage the children to use their knowledge of inverses to work backwards from the answer.

My way

- **Write the answers.**
- **Show how you worked them out on the notepad.**

Don't let your partner see yet!

My method

1.   25 x 13 = _____

2.   7003 − 6995 = _____

3.   30 x 6 = _____

4.   35 x 6 = _____

5.   107 ÷ 32 = _____

6.   1035 + 2993 = _____

7. $\frac{1}{4}$ of 198 = _____

- **For each question, compare your method with a partner's.**
- **Which do you think is easier?**

Now try this!

- **Try the activity again, but this time write your own questions.**

Teachers' note The questions can be masked before photocopying, and others inserted, to create a flexible resource. During the plenary, discuss the range of methods that were used for each question.

**Developing Numeracy
Solving Problems Year 5
© A & C Black**

Telling stories

- **Make up a number story for each number statement.**

  Use g and kg , cm, m and km , ml and l **or** £ and p .

  Example: 633 ÷ 6 = 105·5   6 equal pieces are cut from 633 cm of rope. Each piece is 105·5 cm long.

1. 1243 + 3640 = 4883

_____

_____

_____

_____

2. 53·25 x 24 = 1278

_____

_____

_____

_____

3. 587 – 175 = 412

_____

_____

_____

_____

4. 1345 – 299 = 1046

_____

_____

_____

_____

5. 172 ÷ 8 = 21·5

_____

_____

_____

_____

6. 23·2 x 15 = 348

_____

_____

_____

_____

- **Write three more number statements.**
- **Ask a partner to write matching number stories.**

Teachers' note The instruction, worked example and one number in each statement can be masked before photocopying to create missing number statements. The number stories can then be written as number questions for a partner to solve. You could provide the children with further practice in making up number stories using the 'Spiders' activity on page 5.

Developing Numeracy
Solving Problems Year 5
© A & C Black

On target

- **Use these numbers and**
  $\boxed{+}$, $\boxed{-}$, $\boxed{\times}$ **or** $\boxed{\div}$ **to make the target number.**

You can use a number more than once, or not at all.

| 100 | 10 | 5 | 8 | 3 | 1 |

1. 881 $(100 + 10) \times 8 + 1$

2. 303

3. 153

4. 147

5. 557

| 50 | 10 | 5 | 3 | 2 | 1 |

6. 181

7. 266

8. 132

9. 120

10. 351

Now try this!

- **Use the second set of numbers to make four more target numbers. Give them to your group to solve.**

Teachers' note Practise this type of activity during the oral/mental starter and make sure that the children understand the rules. The children could use calculators where necessary or for checking. Discuss the variety of solutions at the end of the lesson. The activity can be made more demanding by allowing the children to use each number only once.

Developing Numeracy
Solving Problems Year 5
© A & C Black

On target

- **Use these numbers and** $+$, $-$, $\times$ **or** $\div$ **to make the target number.**

You can use each number only once.

50	10	5	8	2	1

1. 794 $(8 \times 2 \times 50) - 5 - 1$

2. 542

3. 408

4. 186

5. 851

75	10	6	9	4	1

6. 757

7. 361

8. 109

9. 397

10. 456

Now try this!

- **Use the second set of numbers to make four more target numbers. Give them to your group to solve.**

Teachers' note Practise this type of activity during the oral/mental starter and make sure that the children understand the rules. The children could use calculators where necessary or for checking. Discuss the variety of solutions at the end of the lesson.

Developing Numeracy
Solving Problems Year 5
© A & C Black

11

Zip Zap puzzles: 1

This grid shows six 3-digit numbers.
They are made from the digits 1 to 9 .

Read '**Zap**' numbers across.
Example: **Zap 1** is 796.

Read '**Zip**' numbers down.
Example: **Zip 1** is 731.

	Zip 1↓	Zip 2↓	Zip 3↓
Zap 1→	7	9	6
Zap 2→	3	2	8
Zap 3→	1	4	5

• **Describe each number using some of these words.**

multiple	even	odd	greater than	less than	between

Zap 1 _is a multiple of 4 less than 800._

Zap 2 _____

Zap 3 _____

Zip 1 _____

Zip 2 _____

Zip 3 _____

• **Read the clues.**

• **Write all the digits** 1 to 9 **on the grid.**

ACROSS
Zap 1 is an even number between 800 and 820.
Zap 2 is an odd number. The sum of the first two
  digits equals the last digit.
Zap 3 is the largest number. It is a multiple of 5.

DOWN
Zip 1 is an odd number between 840 and 860.
Zip 2 is an even number less than 200.
Zip 3 is a multiple of 5 smaller than 300.

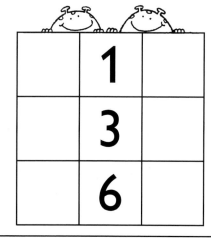

Now try this!

• **Make up your own Zip Zap grid.**

• **Write a clue for each number.**

• **Give it to a partner to solve.**

Use all the
digits 1 to 9.

Teachers' note Discuss that a zip goes up and down, for example on a coat, and that when a
spaceship 'zaps' an alien it shoots across. This can help the children to remember which is which.
The children should use trial and error to find which numbers fit both the Zip and Zap clues. The
following page is a further extension of the work on this page.

Developing Numeracy
Solving Problems Year 5
© A & C Black

Zip Zap puzzles: 2

- **Read the clues.**
- **Complete the grids.**

> Use all the digits 1 to 9.

1. **ACROSS**
Zap 1 is a multiple of 5 between 380 and 390.
Zap 2 is a multiple of 3 between 200 and 300.
Zap 3 is an odd number between 640 and 650.

DOWN
Zip 1 is an even number between 320 and 327.
Zip 2 is the largest number. It is less than 899.
Zip 3 is an odd number less than 520.

2. **ACROSS**
Zap 1 is an odd number less than 130.
Zap 2 is an odd number whose digits total 16.
Zap 3 is a multiple of 5 between 460 and 490.

DOWN
Zip 1 is an even number less than 200. The sum of the first two digits equals the last digit.
Zip 2 is an even number less than 300. The sum of the first two digits equals the last digit.
Zip 3 is the largest number. It is a multiple of 25 between 950 and 999.

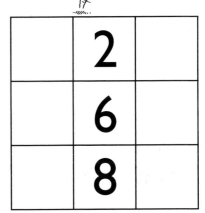

3. **ACROSS**
Zap 1 is a multiple of 25. The first digit equals the sum of the second two digits.
Zap 2 is an even number, just less than 150.
Zap 3 is a number whose digits are three consecutive multiples of 3.

DOWN
Zip 1 is an odd number between 700 and 800.
Zip 2 is less than 300. Its digits are three consecutive multiples of 2.
Zip 3 is an odd number. It is a multiple of 19.

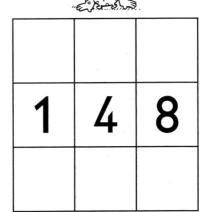

Teachers' note The children should first complete the activity 'Zip Zap puzzles: 1' on page 12. They can use calculators to check which numbers are multiples of a particular large number, such as 19.

Developing Numeracy Solving Problems Year 5 © A & C Black

Pocket the most!

- **You have been given a choice about your pocket money.**

One amount
£2.00 for a week

or

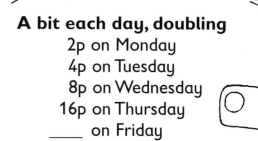

A bit each day, doubling
2p on Monday
4p on Tuesday
8p on Wednesday
16p on Thursday
____ on Friday
____ on Saturday
____ on Sunday

1. Fill in the doubling pattern.
How much would you get in
total for one week? _____

2. Which way gives you the most money in one week?

_____

- **Tick which way gives you the most money in one week.**

3.

One amount
£1.50 for a week

or

A bit each day, doubling
1p on Monday
2p on Tuesday
4p on Wednesday
____ on Thursday
____ on Friday
____ on Saturday
____ on Sunday

4.

One amount
£3.00 for a week

or

A bit each day, doubling
3p on Monday
6p on Tuesday
12p on Wednesday
____ on Thursday
____ on Friday
____ on Saturday
____ on Sunday

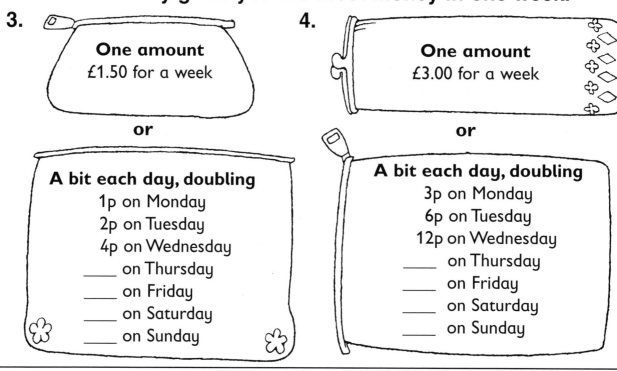

- **This choice is for two weeks. Tick which gives more.**

Now try this!

One amount
£50 for two weeks

or

A bit each day, doubling
1p on Monday
2p on Tuesday
4p on Wednesday...

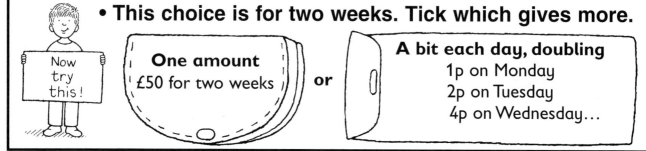

Teachers' note Revise doubling numbers during the first part of the lesson. For further extension, the children could make up their own pocket money puzzles of this type. Some children could be given calculators to use for this investigation.

**Developing Numeracy
Solving Problems Year 5
© A & C Black**

Split it

- **Join 10 cubes in a rod, like this.**

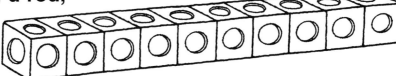

- **Split the rod into two parts. Find the** |product| **of the two numbers.**

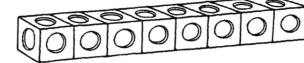

Example: 2   x   8   **product = 16**

---

1. Write all the products you can make by splitting the rod into two parts in different ways.

$1 \times 9 = 9$

$2 \times 8 = 16$

- **Now split the rod into three parts.**

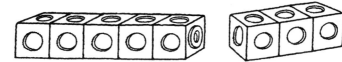

Example: 2   x   5   x   3   **product = 30**

---

2. Write all the products you can make by splitting the rod into three parts in different ways.

Now try this!

- **Try splitting the rod into four parts.**
- **Write all the** |products| **you can make.**

Teachers' note Encourage the children to be systematic when splitting the rod of cubes, for example, splitting off one cube first, then two, and so on. It may be useful to begin this lesson with some oral multiplication questions, including those with three numbers, for example, 2 x 3 x 5. If necessary, revise the term 'product'.

Developing Numeracy Solving Problems Year 5 © A & C Black

Ray the Riddler

- **Solve Ray the Riddler's puzzles.**
- **Write the answers in the bubbles.**

1. Find two consecutive numbers with a total of 187.

__93__   __94__

2. Find two consecutive numbers with a total of 35.

____   ____

3. Find three consecutive numbers with a total of 216.

____   ____   ____

4. Find three consecutive numbers with a total of 108.

____   ____   ____

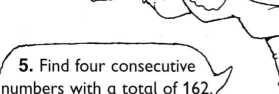

6. Find four consecutive numbers with a total of 398.

____   ____   ____   ____

5. Find four consecutive numbers with a total of 162.

____   ____   ____   ____

8. Find two consecutive numbers with a product of 182.

____   ____

7. Find two consecutive numbers with a product of 132.

____   ____

9. Find two consecutive numbers with a product of 600.

____   ____

10. Find three consecutive numbers with a product of 990.

____   ____   ____

Now try this!

- **Write two more puzzles about consecutive numbers.**
- **Give them to a partner to solve.**

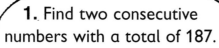

Teachers' note If necessary, remind the children of the meaning of the words 'consecutive' and 'product'. Calculators could be used for this activity. For puzzles with 'products', you could introduce using the square root key on a calculator to find an approximate answer.

Developing Numeracy Solving Problems Year 5 © A & C Black

Sum game!

• Play this game with a partner.

☆ One player uses the numbers 1, 3 and 5.
The other uses the numbers 2, 4 and 6.

☆ Take turns to write a number on the grid.

☆ Score a point if the numbers in a line
add up to a multiple of 7.

☆ The lines can be vertical, horizontal
or diagonal.

☆ The winner is the one with the most points
when the grid is full!

		3	
2	4		1

If you write a 4 here, you score 2 points.
1 point for the vertical line (3 + 4 = 7) and
1 point for the horizontal line (2 + 4 + 1 = 7).

• Play four times. Who scores the most?

• How can you make ⬛14⬛ with two odd numbers
and two even numbers? Write the ways.

Teachers' note Different numbers can be used for this game, making sure one child uses even numbers and one uses odd numbers. Some children may find it easier to play the game looking for multiples of 5, using the odd numbers 1 and 3, and the even numbers 2 and 4. For the extension activity, encourage the children to find the solutions systematically.

Developing Numeracy
Solving Problems Year 5
© A & C Black

Darting about

You score $\boxed{\text{double}}$ points if you hit the shaded part of the dartboard. You score $\boxed{\text{single}}$ points if you hit anywhere else.

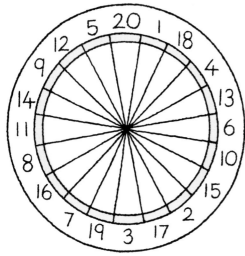

• **You have two darts. Write four ways you can score:** $\boxed{20}$

double 8 + single 4 _____

_____ _____

$\boxed{41}$

_____ _____

$\boxed{52}$

_____ _____

_____ _____

• **You have two darts. If one hits a double section and one hits a single section, how can you make these scores?**

• **Fill in the charts.**

Score	Double	Single
3		
4	1	2
5		
10		
11		
12		
24		
25		
26		
38		
39		

Score	Double	Single
40		
46		
47		
48		
50		
53	20	13
54		
57		
58		
59		
60		

Now try this!

• **You have** $\boxed{\text{three}}$ **darts. What is the highest total you can make on this dartboard?** _____

Teachers' note Begin the lesson by revising doubles to 20. Include questions that involve doubling and then adding a single number. Discuss how a dartboard works for any children unfamiliar with this context.

Developing Numeracy
Solving Problems Year 5
© A & C Black

Darting about

You score ⌈double⌉ points if you hit the outer ring of the dartboard.

You score ⌈triple⌉ points if you hit the inner ring.

You score ⌈single⌉ points if you hit anywhere else.

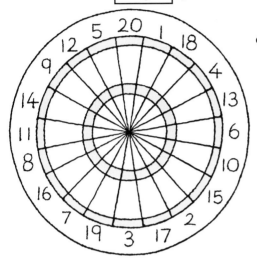

• You have three darts. Write four ways you can score ⌈10⌉.

<u>triple 1 + double 2 + single 3</u>

_____

_____

_____

• Using all three darts, which of these scores can you make?
• Fill in the chart.

Score	Method
77	triple 20, double 5, single 7
87	
97	
107	
117	
127	
137	
147	
157	
167	

• You have three darts. What is the highest total you can make on this dartboard? _____

Teachers' note Begin the lesson by revising doubles to 20 and discussing how to multiply numbers by 3. Discuss the word 'triple' and practise questions that involve doubling and trebling and finding the sum of the answers. Discuss how a dartboard works for any children unfamiliar with this context.

Developing Numeracy
Solving Problems Year 5
© A & C Black

Sum of the digits

- **Write the multiples in order on the trains.**
- **On the wheel under each multiple, write the sum of its digits.**

  Example: 12 → 1 + 2 = 3

  If you get a two-digit answer, do the same again until you reach a single digit. Example: 39 → 3 + 9 = 12 → 1 + 2 = 3

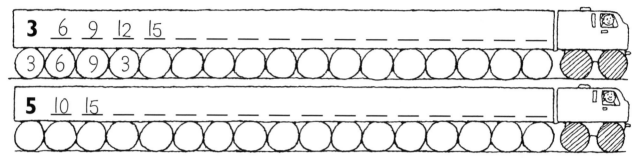

| 3 | 6 | 9 | 12 | 15 | __ __ __ __ __ __ __ __ __ __ |

Wheels: 3 6 9 3 ...

| 5 | 10 | 15 | __ __ __ __ __ __ __ __ __ __ |

- **Using the numbers on the wheels in order (3, 6, 9, 3...), draw lines on the grid. Turn the paper clockwise through** 90° **between each line.**

  Complete the pattern for 3 **.   Complete the pattern for** 5 **.**

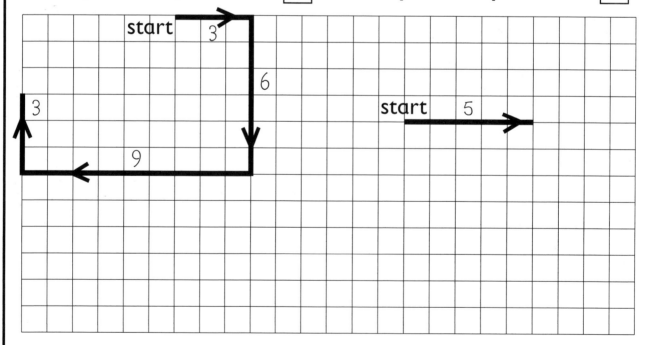

- **Draw a train for** multiples of 8 **.**
- **Draw the pattern for** 8 **on squared paper. Start at the top, in the middle.** ——→ 8

Teachers' note Discuss how to find the sum of the digits, including those that initially give a two-digit answer, for example, 48 → 4 + 8 = 12 → 1 + 2 = 3. This is sometimes known as finding the digital root. Other multiples, for example, of 6 and 9, can be explored in the same way. You could revise multiples with the children using the 'Counting stick' activity on page 5.

**Developing Numeracy
Solving Problems Year 5
© A & C Black**

Fishing for facts

- **Tick** ⬚ true **or** ⬚ false **for each statement.**
- **Write four examples to show why.**

1. Every multiple of 4 is twice a multiple of 2.

true ☐ **false** ☐

$12 = 2 \times 6$

↑ a multiple of 4   ↑ a multiple of 2

2. Every multiple of 6 is twice a multiple of 2.

true ☐ **false** ☐

3. Every multiple of 6 is twice a multiple of 3.

true ☐ **false** ☐

4. Every multiple of 12 is three times a multiple of 4.

true ☐ **false** ☐

Now try this!

- **Write another statement which is true.**
- **Write four examples to prove it.**

Teachers' note If necessary, revise the term 'multiple' and identify some multiples on a number line. Encourage the children to test the statements using some small and some very large numbers. Ensure that the children realise that giving one example is insufficient to prove a general statement, although one example can be enough to disprove it.

**Developing Numeracy
Solving Problems Year 5
© A & C Black**

What's the link?

These questions are all linked. They can be worked out in similar ways.

> How many months are there in 8 years?

> How many months are there in 27 years?

> How many months are there in 100 years?

1. Explain in words how to find the number of months in any number of years. _____

_____

These questions are all linked, too.

> How much change will I get from 50p if I buy 2 chews costing 4p each?

> How much change will I get from 50p if I buy 6 chews costing 4p each?

> How much change will I get from 50p if I buy 10 chews costing 4p each?

2. Explain in words how to find the change from 50p for any number of chews costing 4p each. _____

_____

_____

3. Explain in words how to find the price of any number of CDs costing £8 each. _____

_____

_____

• **Explain in words how to find the number of** ⟨minutes⟩ **in any number of** ⟨days⟩ **.**

Teachers' note The children may require further practice of generalising and describing relationships like these in words. Encourage them to include as much detail as they can in their explanations. Some children could be encouraged to write a formula using letters and numbers, if appropriate.

**Developing Numeracy
Solving Problems Year 5
© A & C Black**

Multiple mayhem

These numbers are all multiples of 9 .

• **Can you see what they have in common?**

Try adding the digits!

405   27   11 511   333   7 101   531   2 223   108   72

1. Explain in words how you can recognise a multiple of 9.

_____

These numbers are all multiples of 90 .

• **Can you see what they have in common?**

450   810   11 520   360   7 110   5 310   2 250   180   720

2. Explain in words how you can recognise a multiple of 90.

_____

Now try this!

• **Write some** multiples of 900 . **Explain in words how you can recognise a** multiple of 900 .

• **Do the same for** multiples of 9 000 .

Teachers' note This work encourages the children to generalise. Ask further questions, such as 'How can you spot a multiple of 2, 20, 200? 5, 50, 500?' etc. Invite the children to explain in words how they can recognise these multiples.

**Developing Numeracy
Solving Problems Year 5
© A & C Black**

Spot the pattern

- **Continue the sequences.**
- **Explain the patterns in words.**

1.

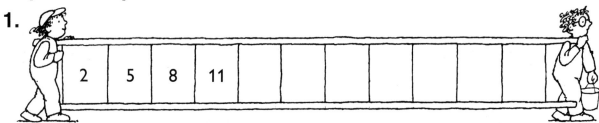

| 2 | 5 | 8 | 11 | | | | | | | | | |

Pattern: _____

2.

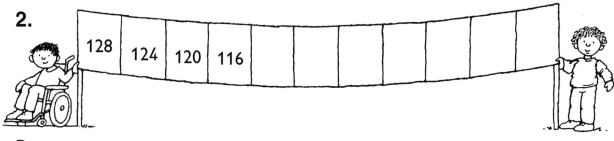

| 128 | 124 | 120 | 116 | | | | | | | |

Pattern: _____

3.

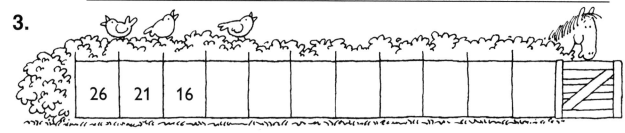

| 26 | 21 | 16 | | | | | | | | | | |

Pattern: _____

4.

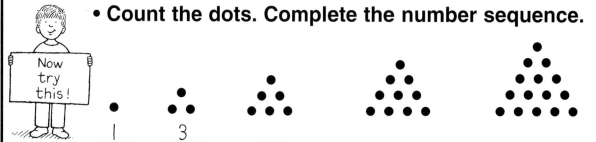

| 1 | 4 | 9 | 16 | 25 | | | | | |

Pattern: _____

- **Count the dots. Complete the number sequence.**

Now try this!

1   3   ___   ___   ___

- **What are these numbers called?** _____

Teachers' note Practise counting forwards and backwards during the oral/mental starter.
Encourage the children to find the difference between adjacent numbers in a sequence and
to look for patterns in these numbers. The numbers in the sequences could be masked before
photocopying, and others inserted, to create a flexible resource.

**Developing Numeracy
Solving Problems Year 5**
© A & C Black

24

- **Continue the sequences.**

- **Explain the patterns in words.**

1.

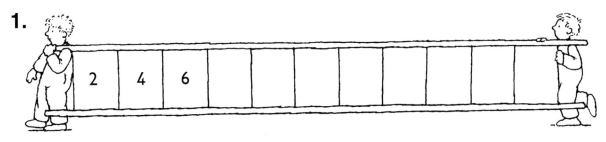

| 2 | 4 | 6 | | | | | | | |

Pattern: _____

What will the 20th number in the sequence be? _____

2.

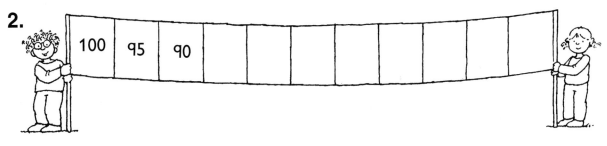

| 100 | 95 | 90 | | | | | | |

Pattern: _____

What will the 20th number in the sequence be? _____

3.

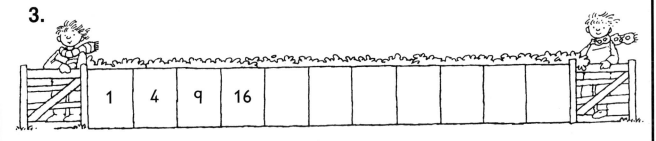

| 1 | 4 | 9 | 16 | | | | | | |

Pattern: _____

What will the 20th number in the sequence be? _____

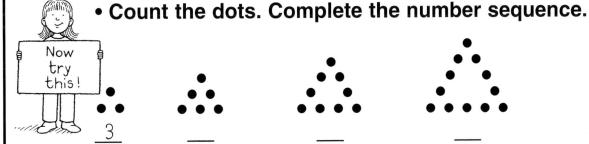

- **Count the dots. Complete the number sequence.**

Now
try
this!

_3_ ___ ___ ___

- **What will the** [10th] **number in the sequence be?** _____

Teachers' note Practise counting forwards and backwards during the oral/mental starter.
Encourage the children to look for a relationship between the number and its position in the
sequence (sometimes called its 'term' number). The numbers in the sequences could be masked
before photocopying, and others inserted, to create a flexible resource.

**Developing Numeracy
Solving Problems Year 5**
© A & C Black

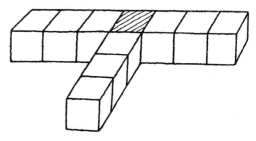

• **Make these models using cubes.**

model 1   model 2   model 3

1. How many cubes would you need to make

model 4? _____   model 6? _____   model 10? _____

• **Make these models using cubes.**

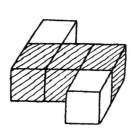   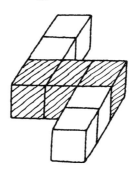   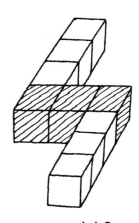

model 1   model 2   model 3

2. How many cubes would you need to make

model 4? _____   model 6? _____   model 10? _____

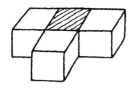 Now try this!

• <u>Without</u> making these models, describe the pattern in words.

          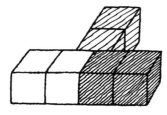   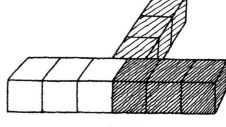

model 1   model 2   model 3

• **How many cubes would you need to make**

model 6? _____   model 10? _____   model 100? _____

Teachers' note Practise counting forwards and backwards during the oral/mental starter.
Encourage the children to look for a relationship between the number and its position in the
sequence (sometimes called its 'term' number).

**Developing Numeracy
Solving Problems Year 5**
© A & C Black

Peaches

This tray has space for six peaches.

Here is one peach.

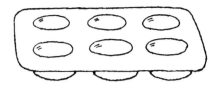

There are six ways you can put one peach in the tray.

- **Record the ways.**

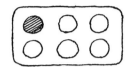 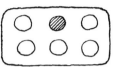 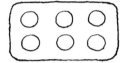

 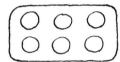 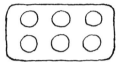

- **How can you arrange two peaches in the tray? Record the ways.**

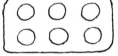 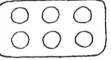 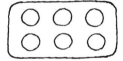

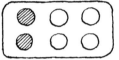  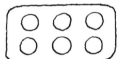  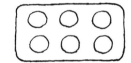

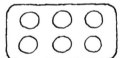  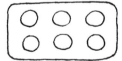  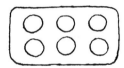

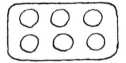 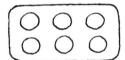

Now try this!

- **Work out how many ways there are of arranging** three **and** five **peaches in the tray.**
- **Complete the chart.**

Number of peaches	1	2	3	4	5
Number of arrangements	6	15		15	

- **What pattern do you notice?**

Teachers' note Encourage the children to compare results to ensure that the group finds all the solutions. Provide squared paper for the extension activity.

Developing Numeracy
Solving Problems Year 5
© A & C Black

Perfect picnic

**Jack and Molly are making sandwiches for a picnic.
They decide to make each one different.**

We have brown bread, white bread...

...jam, chocolate spread and cheese.

1. They use only one filling in each sandwich. Write all the

different sandwiches they can make.

brown bread and jam

2. How many different sandwiches can they make? _____

Molly's dad arrives with some wholemeal bread.

3. How many <u>more</u> sandwiches can they make? _____

4. How many different sandwiches can they make in total now? _____

Now try this!

**Molly wants to make a sandwich with
two fillings.**

- **Write all the sandwiches she could
  make with the three types of bread.**

Teachers' note At the end of the lesson, show the children the different combinations by drawing a table with the types of bread down the side and the fillings across the top. You could then ask, 'How many sandwiches can be made in this way if there were four types of bread/an extra filling?'

**Developing Numeracy
Solving Problems Year 5
© A & C Black**

Count them up

• **Count all the** `triangles` **in each shape.**

1.

Be careful!
There are triangles
within triangles.

_____ triangles

2.

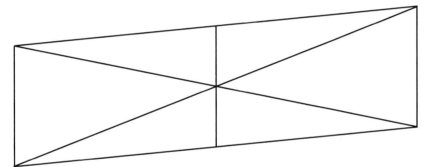

_____ triangles

3.

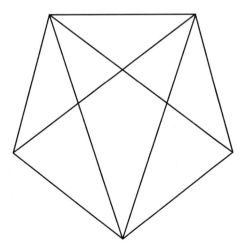

_____ triangles

`Now try this!`

• **Cut a 3 x 3 grid from squared paper.**

• **How many** `squares` **can you see?** _____

• **How many** `rectangles` **can you see?** _____

Teachers' note Encourage the children to be systematic. For the third shape, ask them to count all the triangles made from one piece, then the triangles made from two, then three, etc. They should be aware that one of the middle sections is not a triangle.

**Developing Numeracy
Solving Problems Year 5
© A & C Black**

Shape puzzles

A │pentomino│ is a shape made from five squares that touch along at least one edge.

Examples:

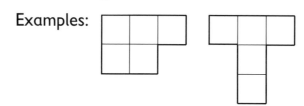

1. Draw as many different pentominoes as you can.

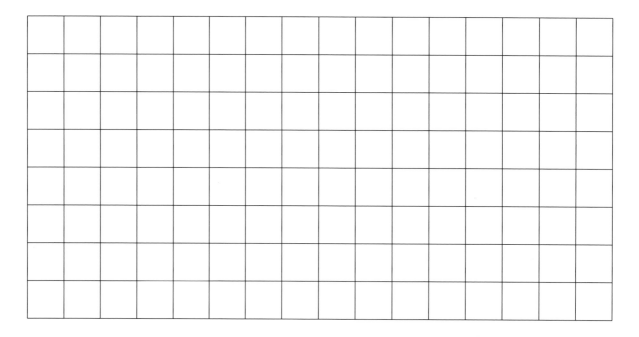

2. Find different ways to divide this shape into two pentominoes. Shade the two pentominoes in different colours.

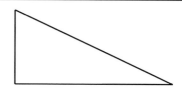

Draw more ways on squared paper.

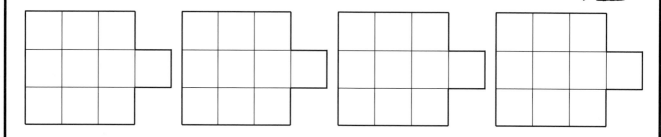

Now try this!

• **If this is │half│ a shape, what could the whole shape be? Draw as many shapes as you can.**

Teachers' note Demonstrate to the children that although a mirror image pentomino may look different, it is in fact the same shape. For the extension activity, encourage the children to sketch the different possibilities. They could choose a different half shape to explore in the same way, for example, a parallelogram, a rectangle, or a triangle which is half a square.

Developing Numeracy Solving Problems Year 5 © A & C Black

Shape puzzles

A hexomino **is a shape made from six squares that touch along at least one edge.**

Examples: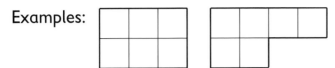

1. Draw as many different hexominoes as you can.

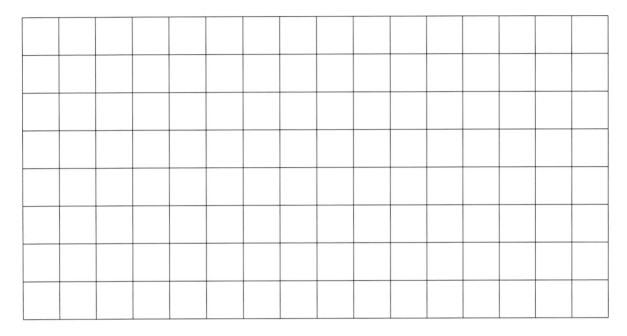

2. Find different ways to divide this shape into two hexominoes. Shade the two hexominoes in different colours.

Draw more ways on squared paper.

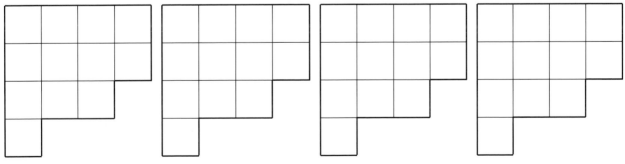

Now try this!

- **If this is** half **a shape, what could the whole shape be? Draw as many shapes as you can.**

Teachers' note Demonstrate to the children that although a mirror image hexomino may look different, it is in fact the same shape. For the extension activity, encourage the children to sketch the different possibilities. They could choose a different half shape to explore in the same way, for example, a parallelogram, a rectangle, or a triangle which is half a square.

**Developing Numeracy
Solving Problems Year 5
© A & C Black**

These dots are the [vertices] of three squares.

• **Draw the squares.**

These dots are the [vertices] of three regular hexagons.

• **Draw the hexagons.**

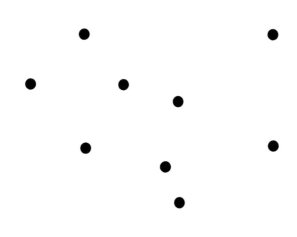

• **Make your own hidden shape puzzle. Choose some shape templates and draw dots at the vertices.**

• **Give your puzzle to a partner to solve.**

Teachers' note Encourage the children to use a ruler to measure the lines and to ensure that they draw regular shapes. If necessary, revise the terms 'vertices'.

Developing Numeracy
Solving Problems Year 5
© A & C Black

Hidden shapes

These dots are the | vertices | of four squares.

• Draw the squares.

These dots are the | vertices | of five equilateral triangles.

• Draw the triangles.

• **Make your own hidden shape puzzle. Choose some shape templates and draw dots at the vertices.**

• **Give your puzzle to a partner to solve.**

Teachers' note Encourage the children to use a ruler to measure the lines and to ensure that they draw regular shapes.

Developing Numeracy
Solving Problems Year 5
© A & C Black

Amazing angles

Josh is measuring angles with a protractor.

When I measure two angles on a straight line, the angles always add up to 180°.

● **Measure these angles to see whether Josh is right.**

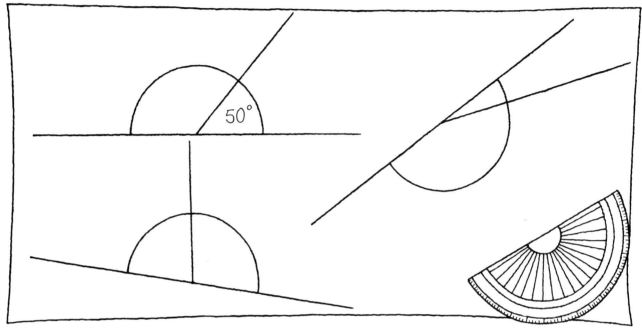

● **Draw some more angles on a straight line to test whether this always works.**

● **Do you think that every pair of angles on a straight line adds up to** `180°` **? _____**

Teachers' note The children should be encouraged to draw pairs of angles in different orientations and with sides of different lengths. If necessary, remind the children how to use a protractor accurately. Some children could measure the angles to the nearest five degrees.

**Developing Numeracy
Solving Problems Year 5
© A & C Black**

• **Use shape templates to draw these shapes.**

| equilateral triangle | | square | | regular pentagon | | regular hexagon |

| regular octagon |

1. Measure the length of one side of each shape. Write it in the chart.

2. Without measuring, work out the perimeter of each shape. Complete the chart.

Shape	Length of one side	Perimeter
equilateral triangle		
square		
regular pentagon		
regular hexagon		
regular octagon		

3. Explain how to find the perimeter of any regular shape.

_____

_____

Teachers' note Explain to the children that the shapes can overlap if necessary. Encourage them to generalise how they found the perimeters and to write an explanation with plenty of detail that would apply to any regular shape, for example a regular 100-sided shape! The children could also use a simple formula to describe this.

**Developing Numeracy
Solving Problems Year 5
© A & C Black**

School survey

There are _____ books in our library.

There are _____ children in our class.

There are _____ classes in our school.

There are _____ children in our school.

- **Use this information to answer the questions.**

1. If each child in our class takes out a book from the library, how many books will be left? _____

2. What is the difference between the number of books in the library and the number of children in the school? _____

3. If all the books in the library are packed equally into ten boxes, how many books will be in each box? _____

4. If all the classes in the school had about the same number of children, how many would this be? _____

5. Is your class larger than this? _____

6. If the library books are shared out between the classes, how many books will each class get? _____

7. How many books are there for each child in the school? _____

Now try this!

- **How many library books will be left if each child in your class takes out**
2 books? _____ 5 books? _____

Teachers' note Before giving this sheet to the children, fill in the statistics at the top of the sheet. Alternatively, as a preparatory activity, ask the children to find out this information for themselves, for example, they could find out the number of books in the library by estimating the number on one shelf and multiplying this by the number of shelves.

**Developing Numeracy
Solving Problems Year 5**
© A & C Black

Rapid readers

• **Answer these questions.**

In my book, the story starts on page 1 and is 510 pages long. I have read 134 pages.

1. How many pages have I still to read? _____

2. How many more pages must I read to reach the middle of the book? _____

3. If the book has 10 equal chapters, which chapter am I reading? _____

4. On which page does chapter 9 start? _____

5. I usually read about 20 pages a day. For how many days have I been reading the book? _____

6. How many more days do you think it will take me to finish the book? _____

In my book, the story starts on page 1 and is 350 pages long. I have read 78 pages.

7. How many pages have I still to read? _____

8. How many more pages must I read to reach the middle of the book? _____

9. If the book has 10 equal chapters, which chapter am I reading? _____

10. On which page does chapter 4 start? _____

Now try this!

• **It took me 2 days to read** 126 pages **. How long might it take me to read** 440 pages **?** _____

Teachers' note Remind the children that they may give approximate answers where appropriate. Further activities on this theme could involve finding the number of pages read at a particular rate in January, February, etc.

Developing Numeracy
Solving Problems Year 5
© A & C Black

Going loopy

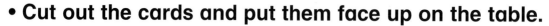

- **Cut out the cards and put them face up on the table.**
- **Pick a card and answer the question.**
- **Find the card that starts with that answer. Place the cards in a loop.**

36 eggs
17 get broken
How many eggs?

2 full boxes of eggs
6 in a box
1 box gets broken
How many eggs?

8 full boxes of eggs
6 eggs in each box
How many eggs?

4 egg boxes, 3 full
1 egg in the 4th box
2 eggs get broken
How many eggs?

No eggs
48 eggs bought
6 eggs in a box
How many full boxes?

3 eggs
3 get broken
How many eggs?

19 eggs
6 eggs per box
How many boxes needed?
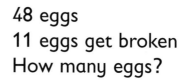

48 eggs
11 eggs get broken
How many eggs?

6 eggs
Half of the eggs break
How many eggs?

6 full boxes of eggs
6 eggs in a box
How many eggs?

17 eggs
6 eggs per box
How many full boxes?

37 eggs
6 eggs per box
How many full boxes?

Teachers' note The cards form one continual loop. They can be used for a whole class or group activity, where each child has a card and the questions are read aloud.

**Developing Numeracy
Solving Problems Year 5
© A & C Black**

Saving up

Amy saves $\boxed{£1.50}$ **of her pocket money every week.**

1. How much does she save:

in 4 weeks?  £6   in 10 weeks?

Amy starts saving $\boxed{£2.00}$ **a week.**

2. How much <u>extra</u> does she save:

in 4 weeks?    in 10 weeks?

3. Amy saves £1.50 for 10 weeks and £2.00 for 4 weeks.

How much does she save?

4. Amy spends £3 on a new book, twice as much on a video and 99p on a big bar of chocolate.

How much of her savings does she have now?

Javed saves $\boxed{£1.20}$ **a week for a whole year.**

5. How much does he save:

in 4 weeks?    in 10 weeks?

in half a year?    in a whole year?

 • **If Javed's mum saves** $\boxed{£150}$ **a month, how much does she save in** $\boxed{\text{2 months?}}$ $\boxed{\text{1 year?}}$ $\boxed{\text{1 decade?}}$

Teachers' note As an oral/mental starter, remind the children of the meaning of the word 'decade' and revise the number of months in a year, weeks in a month/year, etc.

**Developing Numeracy
Solving Problems Year 5
© A & C Black**

39

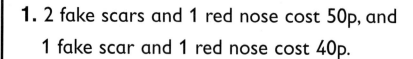

• **Solve these joke shop problems.**

1. 2 fake scars and 1 red nose cost 50p, and
1 fake scar and 1 red nose cost 40p.

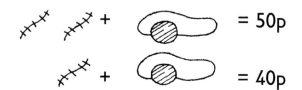 = 50p

+ = 40p

To find out how much
a red nose costs, first work
out how much a scar costs.

A scar costs _____ . A red nose costs _____ .

2. 2 spiders and 1 snake cost 60p, and
1 spider and 1 snake cost 45p.

+ = 60p

+ = 45p

A spider costs _____ . A snake costs _____ .

3. 2 bats and 1 set of vampire teeth cost 95p, and
1 bat and 1 set of vampire teeth cost 65p.

+ = 95p

+ = 65p

A bat costs _____ . A set of vampire teeth costs _____ .

4. 1 mask and 2 hairy hands cost 88p, and
1 mask and 1 hairy hand cost 57p.

A hairy hand costs _____ .

A mask costs _____ .

Teachers' note Discuss how subtraction of the two prices can be used to find the price of the
missing item from the second 'equation' in each pair. For practice in adding prices, you could use
the 'Shopping trip' activity on page 5.

**Developing Numeracy
Solving Problems Year 5
© A & C Black**

• **Work out how much each item costs.**

1. 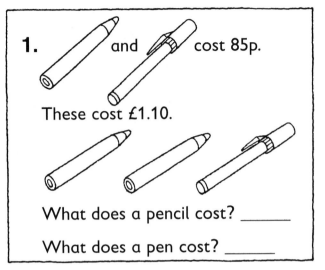 and _____ cost 85p.

These cost £1.10.

What does a pencil cost? _____

What does a pen cost? _____

2. 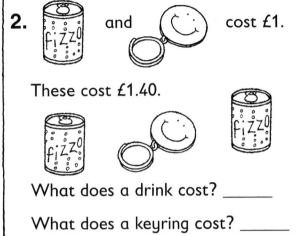 and _____ cost £1.

These cost £1.40.

What does a drink cost? _____

What does a keyring cost? _____

3. 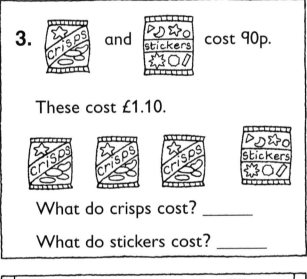 and _____ cost 90p.

These cost £1.10.

What do crisps cost? _____

What do stickers cost? _____

4. These cost £1.40.

These cost 95p.

These cost 50p.

What does a badge cost? _____

What does a sweet cost? _____

What does a comic cost? _____

Mr Toffee's Sweet Shop No 9

BUY BAZO TODAY

fizzled out try fizzo!

open

Now try this!

• **Choose three items of your own.**

• **Make up a price for each.**

• **Write a price puzzle for your items.**

Teachers' note Discuss how subtraction of the two prices can be used to find the price of the missing item from the second 'equation' in each pair. The children should be systematic when tackling the extension activity. Encourage them to suggest realistic possibilities.

**Developing Numeracy
Solving Problems Year 5
© A & C Black**

Football mad

John is football mad!

1. Last year John saw 40 matches. If each ticket cost £16, how much did he spend? _____

2. Last week John and three of his friends spent £72 on tickets for a match. How much was each ticket? _____

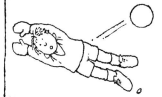

3. At each match, John spent £1.50 on a hotdog. How much did he spend at 40 matches? _____

4. John's train ticket cost £5 for each match. How much did he spend for 40 matches? _____

5. A pack of football stickers costs 25p. John buys three packs every week. How much does he spend on stickers in

4 weeks? _____   6 weeks? _____

10 weeks? _____   40 weeks? _____

A football kit costs exactly £75.

• **Write on the labels how much each item in the kit could cost. They are in order of price, starting with the most expensive.**

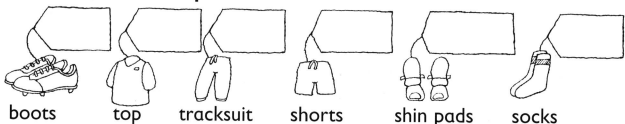

boots   top   tracksuit bottoms   shorts   shin pads   socks

Teachers' note The numbers on the sheet could be masked before photocopying, and others inserted, to create a flexible resource.

**Developing Numeracy
Solving Problems Year 5
© A & C Black**

Bargain buys

• **Work out how much each item costs in the sale.**

1. **A**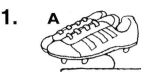
    50% off
    £48
    sale price
    £24

    B
    50% off
    £46
    sale price

    C
    25% off
    £28
    sale price

    D
    10% off
    £20
    sale price

    Which boots are the cheapest? _____

    Which boots are the most expensive? _____

2. **A**
    SAGO TIME RAIDER
    75% off
    £44
    sale price

    B
    ZOOM DINO
    10% off
    £70
    sale price

    C
    VIRTUAL ZAPPER
    90% off
    £120
    sale price

    D
    GAMEGIRL FOOTER
    25% off
    £24
    sale price

    Which game is the cheapest? _____

    Which game is the most expensive? _____

3. **A**
    50% off
    £17
    sale price

    B
    25% off
    £14
    sale price

    C
    75% off
    £18
    sale price

    D
    20% off
    £25
    sale price

    Which jacket is the cheapest? _____

    Which jacket is the most expensive? _____

 Now try this! • **Find 1% of** £500 , £120 , £30 **and** £6 .

Teachers' note Remind the children of quick ways of finding percentages mentally, for example, 25% is half and half again. Discuss the relationship between 25% and 75%. The numbers used on the sheet could be masked and altered before photocopying to provide a more flexible resource to meet the needs of the children.

**Developing Numeracy
Solving Problems Year 5
© A & C Black**

Money exchange

Sam goes on holiday to the USA, where they use dollars.

1. If £1 is about 2 dollars, how much would each item cost in dollars?

| £5 | £16 | £38 | £95 |

| 10 dollars | | | |

Alex goes on holiday to France, where they use francs.

2. If £1 is about 9 francs, how much would each item cost in francs?

| £5 | £7 | £11 | £120 |

| | | | |

Lucy goes on holiday to Greece, where they use drachmas.

3. If £1 is about 500 drachmas, how much would each item cost in drachmas?

| £3 | £8 | £12 | £60 |

| | | | |

 Now try this!

• **I have** £50 **. How many** dollars **,** francs **or** drachmas **could I exchange it for?**

Teachers' note Ensure that the children understand the idea of prices in different currencies being worth the same but with different-sized numbers.

**Developing Numeracy
Solving Problems Year 5
© A & C Black**

Money exchange

Sam goes on holiday to the USA, where they use dollars.

1. If £1 is about 1.6 dollars, how much would each item cost in dollars?

£2   £4   £20   £80

| 3.2 dollars | | | |

Alex goes on holiday to France, where they use francs.

2. If £1 is about 8.7 francs, how much would each item cost in francs?

£2   £5   £25   £120

Lucy goes on holiday to Greece, where they use drachmas.

3. If £1 is about 450 drachmas, how much would each item cost in drachmas?

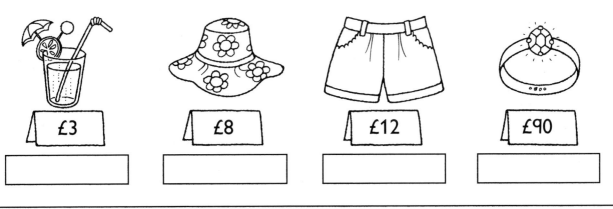

£3   £8   £12   £90

Now try this!

• **I have** £50 . **How many** dollars , francs **or** drachmas **could I exchange it for?**

Teachers' note The answers to the extension activity will be different from those on the previous page. Discuss with the children that these are more precise answers and that the exchange rate numbers on the previous sheet have been rounded to the nearest whole number or 100.

**Developing Numeracy
Solving Problems Year 5**
© A & C Black

Super swimmer

• **Read the newspaper report.**

SUPER SWIMMER LINDA!
Linda Bendall, age 16, swims at the Riverside pool for a staggering $2\frac{1}{2}$ hours a day! Super Linda swims 75 lengths of the pool every day.

'I swim either in the morning or in the afternoon,' says Linda.

• **Answer these questions.**

1. How many hours does Linda swim in

2 days? _____ 1 week? _____ 2 weeks? _____

1 month? _____ 1 year? _____

2. If she starts swimming at 05:55, what time does she finish? _____

3. If she starts swimming at 16:40, what time does she finish? _____

4. Louise swims for half the time that Linda does. For how long does Louise swim? _____

5. Approximately how long does it take Linda to swim one length? _____

6. In a competition, Linda swam for 3 hours and 24 minutes. How much longer was this than her normal swim? _____

• **If the length of the pool is** 50 m **, how far does Linda swim each day?** _____

Teachers' note The following page continues the theme of Linda the Super Swimmer. Discuss that approximate answers could be given for one month and one year, for example, using 4 weeks or 12 months rather than 30/31 days or 365/366 days.

**Developing Numeracy
Solving Problems Year 5**
© A & C Black

Super swimmer grounded!

• **Read the newspaper report.**

LINDA LOSES OUT!

The Riverside pool closes for repairs, leaving Super Swimmer Linda Bendall without a pool to practise in for seven weeks.

'I've been paying £1.20 every day and this is the way they treat me,' says Linda.

• **Answer these questions.**

1. How much money does Linda pay to swim for

2 days? _____ 1 week? _____ 2 weeks? _____

1 month? _____ 1 year? _____

2. If the pool closes on 30 April, on what date should it

open again? _____

3. The repairs take longer than expected. The pool finally opens on

16 July. For how many weeks was it shut? _____

4. On how many days did Linda miss swimming at the pool? _____

5. If Linda normally swims for $2\frac{1}{2}$ hours a day, how many hours of

swimming did she miss? _____

Now try this!

Julie also swims at the Riverside pool. She pays £6.00 **a week.**

• **How many times does Julie go swimming each week?** _____

Teachers' note This page contains a mixture of problems involving money and time. The children may require a calendar to help them with questions 2 and 3. Discuss that approximate answers could be given for one month and one year, for example, using 4 weeks or 12 months rather than 30/31 days or 365/366 days.

**Developing Numeracy
Solving Problems Year 5
© A & C Black**

Wrap it up

Ranjit's family are giving each other presents. There are seven people in the family, so Ranjit is wrapping six presents.

1. Ranjit has 4 m of ribbon. One bow uses 60 cm of ribbon.

Is there enough ribbon to make 6 bows? _____

2. How much ribbon will Ranjit have left over? _____

3. Is this enough to make an extra bow? _____

4. Ranjit uses 26 cm of sticky tape for each present.

How much does he use altogether? _____

5. There was 3 m of sticky tape on the roll before Ranjit wrapped

his presents. How much is there now? _____

6. Ranjit uses another 47 cm of sticky tape putting

up decorations. How much is left on

the roll? _____

7. Ranjit's mum cuts 7 pieces of ribbon

from a 3 m roll. Each piece is 42 cm long.

How much ribbon is left? _____

8. Ranjit's sister sticks ribbon along 4 sides of a

present. The present is a cube with sides of 32 cm.

How much ribbon does she use? _____

• **If everyone in Ranjit's family gives each other a**
present, how many presents are there altogether? _____

Teachers' note A further activity could involve looking at different-shaped and different-sized presents, for example, triangular, hexagonal or octagonal prisms (with regular cross-sections), and working out how much ribbon would be needed to wrap around the sides.

Developing Numeracy
Solving Problems Year 5
© A & C Black

Sign language

• Colour the correct answer.

1. It is 332 miles from Aberdeen to Manchester, and 192 miles from Manchester to London. How far is it from Aberdeen to London via Manchester?

| 524 miles | | 370 miles |
| 114 miles | | 468 miles |

2. A family cover half their journey before lunch and a quarter after lunch. They still have 96 miles to go. How far is the journey?

| 192 miles | 384 miles |
| 48 miles | 484 miles |

3. A lorry driver travels 86 miles before lunch and half as far after lunch. How far does he drive?

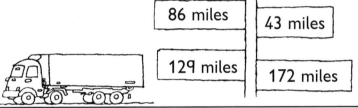

| 86 miles | 43 miles |
| 129 miles | 172 miles |

4. Each side of an octagonal field is 175 m. How much fencing is needed to go around the perimeter?

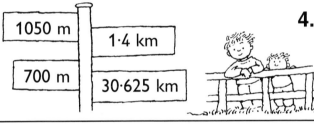

| 1050 m | 1·4 km |
| 700 m | 30·625 km |

5. An ice cream van travels 7·2 km in the morning and 900 m in the afternoon. How far does it travel?

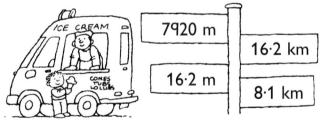

| 7920 m | 16·2 km |
| 16·2 m | 8·1 km |

6. 3600 m of fencing fits exactly around a square field. How long is one side of the field?

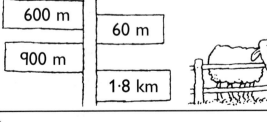

| 600 m | 60 m |
| 900 m | 1·8 km |

Now try this!

• Which was the most difficult to work out? Why?
• Which was the easiest to work out? Why?

Teachers' note This could be carried out as a pairs game, in which each pair is given the same question and must explain their choice of answer. Points can be awarded for correct answers.

Developing Numeracy
Solving Problems Year 5
© A & C Black

Classroom refit

Class 5 are having their room modernised.

1. The length of the classroom is twice its width.
   The width is 5·5 m. How long is the room? _____

2. Each new table top has a perimeter of 3 m. The length of each table is 1 m. What is the width? _____

3. A computer measures 80 cm. It covers half the length of the computer desk. How long is the desk? _____

4. Two shelves are 175 cm and 1·1 m in length. What is the total length of the shelves in centimetres? _____

   in metres? _____

5. What is the difference between the lengths of the shelves in centimetres? _____

   in metres? _____

6. The classroom is 5·5 m wide. It is going to be made wider. When it is finished, there will be room for 8 cupboards side by side. Each cupboard is 90 cm wide. By how much will the room be extended? _____

Now try this!

Three bookcases measure 324 cm **in length altogether.**
Two are identical and one is 84 cm **smaller.**
• **What is the length of each bookcase?**

Teachers' note Remind the children to watch out for measurements given in different units, for example, cm and m. They should be reminded to change one of them so that the same unit is used for both. Before beginning the activity, revise the number of centimetres in a metre.

Developing Numeracy
Solving Problems Year 5
© A & C Black

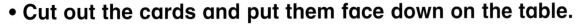

Card competition

- **Cut out the cards and put them face down on the table.**
- **Turn one over. Each player writes down an answer.**
- **Do this for all the cards.**
- **Compare answers. Who got most right?**

Play with a partner.

---

1. 1 kg of potatoes, 300 g of carrots and $\frac{1}{4}$ kg of peas are used in a stew. What weight of vegetables is used?

2. Sarah weighs 38 kg. Pete is 300 g heavier than Sarah. How much does Pete weigh?

3. 6 tennis balls weigh 330 g. How much does 1 ball weigh?

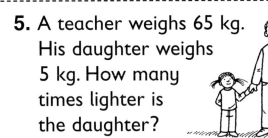

4. A plate weighs 230 g. How much more than 2 kg do 10 plates weigh?

5. A teacher weighs 65 kg. His daughter weighs 5 kg. How many times lighter is the daughter?

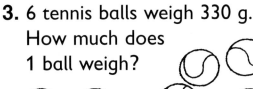

6. Sam buys 640 kg of gravel. He uses a quarter on his drive and the rest on his garden. How much goes on the garden?

7. 6 paving stones weigh 39 kg. How much does 1 paving stone weigh?

8. There is 2·2 kg of sugar in a bag. How much sugar is there in 10 bags?

9. A cat weighs 675 g less than 1 kg. How much does the cat weigh?

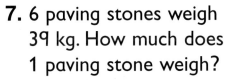

10. A jar of jam weighs 175 g. How much do 5 pots weigh?

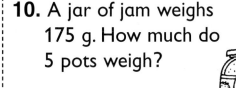

Teachers' note The children could use calculators to check their answers. This activity could be completed as an individual exercise without cutting out the cards. Alternatively, the children could generate their own questions on the backs of the cards.

**Developing Numeracy
Solving Problems Year 5
© A & C Black**

Cookbook quiz

- Work out the total number of grams in each recipe.
- Work out the **approximate** number of grams of each ingredient you would need for one person.

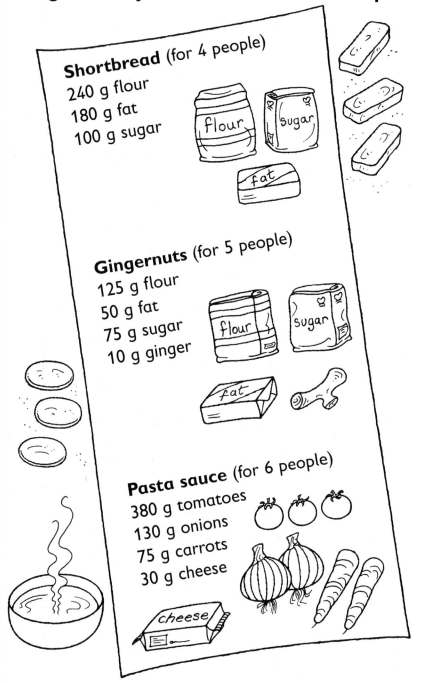

Shortbread (for 4 people)
240 g flour
180 g fat
100 g sugar

Gingernuts (for 5 people)
125 g flour
50 g fat
75 g sugar
10 g ginger

Pasta sauce (for 6 people)
380 g tomatoes
130 g onions
75 g carrots
30 g cheese

Shortbread

total grams _____520 g_____

Recipe for 1 person

flour _____

fat _____

sugar _____

Gingernuts

total grams _____

Recipe for 1 person

flour _____

fat _____

sugar _____

ginger _____

Pasta sauce

total grams _____

Recipe for 1 person

tomatoes _____

onions _____

carrots _____

cheese _____

Now try this!

Here is a recipe for 4 people.
- **Change the recipe to make cookies for 6 people.**

Chocolate cookies
180 g flour
100 g fat
80 g sugar
120 g chocolate

Teachers' note The measurements can be masked before photocopying, and others inserted, to simplify or extend this activity. If necessary, calculators could be used by some children.

**Developing Numeracy
Solving Problems Year 5
© A & C Black**

Fruit and veg

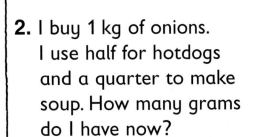

1. I buy 2 kg of carrots and I use 400 g in soup. How many grams do I have now? _____

2. I buy 1 kg of onions. I use half for hotdogs and a quarter to make soup. How many grams do I have now? _____

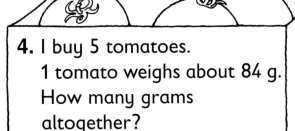

3. A bag of apples weighs 600 g. There are 8 apples in the bag. About how much does each apple weigh? _____

4. I buy 5 tomatoes. 1 tomato weighs about 84 g. How many grams altogether? _____

If I need 500 g of tomatoes, how many more tomatoes do I need to buy? _____

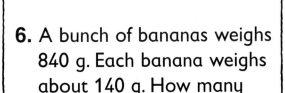

5. A potato weighs about 250 g. About how many potatoes will I get in a

1 kg bag? _____

5 kg bag? _____

10 kg bag? _____

6. A bunch of bananas weighs 840 g. Each banana weighs about 140 g. How many bananas in the bunch?

_____

Now try this!

Each bag of fruit weighs ⟨1 kg⟩.

• **Work out the <u>approximate</u> mass of one item from each bag.**

2 melons   5 apples   7 lemons   12 bananas   14 plums

Teachers' note Reinforce the idea that answers may be approximate and encourage the children to check that their answers are reasonable. Revise the term 'mass' if necessary for the extension activity.

Developing Numeracy
Solving Problems Year 5
© A & C Black

Masterchef

• **Read Billy's recipe for salad dressing.**

400 ml of oil
$\frac{1}{2}$ litre of white wine vinegar
2 ml of diluted mustard
a pinch of salt and pepper

1. How much liquid does Billy make? ml

2. Billy puts the dressing into a litre bottle.
How much more liquid could fit into the bottle? ml

3. A tablespoon holds 20 ml. About how many
tablespoons of dressing are there?

4. A teaspoon holds 5 ml. About how many teaspoons
of dressing are there?

5. Billy's mum uses half the dressing. How much does she use? ml

Julie uses 7 tomatoes to make $\boxed{\frac{1}{4}\text{ litre}}$ **of tomato sauce.**

6. How many tomatoes does she need to make:

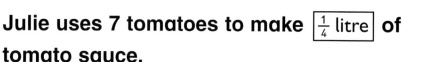

1 litre?   2 litres?

10 litres?   750 ml?

I pour $\boxed{750\text{ ml}}$ **of dressing into two bottles. Both bottles
are full. One holds** $\boxed{\text{half}}$ **as much as the other.**

• **How many millilitres
does each bottle hold?**

Teachers' note The measurements can be masked before photocopying, and others inserted, to simplify or extend this activity. Decimals could be included to provide more of a challenge, for example 1·2 l.

**Developing Numeracy
Solving Problems Year 5
© A & C Black**

Baffling bathrooms

- **Look at this bathroom shelf.**

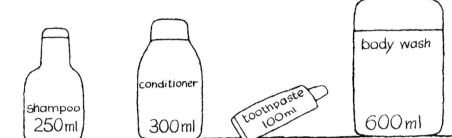

shampoo 250 ml

conditioner 300 ml

toothpaste 100 ml

body wash 600 ml

bubble bath 750 ml

- **Answer the questions.**

1. I use about 18 ml of shampoo every time I wash my hair. About how many washes can I have from the bottle? _____

2. I use half as much conditioner as shampoo every day. How many days will the bottle of conditioner last? _____

3. If I have used 180 ml of bubble bath, how much is left? _____

4. I use half of the toothpaste in 50 days. After how many more days will there be just 20 ml left? _____

5. About how many millilitres of conditioner are there if $\frac{1}{8}$ of it has been used? _____

6. I have 37 showers and use up all the body wash. About how much do I use each time?

_____

Now try this!

I use ⌈1 part⌉ bubble bath to ⌈500 parts⌉ water.

- If I use ⌈90 ml⌉ of bubble bath, how many litres of water do I use? _____

- How many litres of liquid altogether? _____

Teachers' note Discussion between pairs of children is useful here. Alternatively, the children can be encouraged to explain their reasoning to a partner. They could use a calculator to check their answers.

**Developing Numeracy
Solving Problems Year 5
© A & C Black**

Traffic jam

- **Cut out the cards. Put each answer next to the correct question to make a loop.**

A: 43 l

Q: My car holds 43 l of petrol. It is half full. How much petrol is there?

A: 14·5 l

Q: I have 14·5 l of petrol. I put in 28·5 l. How much petrol have I now?

A: 20·4 l

Q: I have 20·4 l of petrol. I use 17·9 l. How much petrol have I now?

A: 2·5 l

Q: I have 2·5 l of petrol. I put in 10 times as much. How much petrol have I now?

A: 37 l

Q: I have 37 l of petrol. I use 22·5 l. How much petrol have I now?

A: 21·5 l

Q: I have 21·5 l of petrol. I use 17·7 l. How much petrol have I now?

A: 6 l

Q: I have 6 l of petrol. I fill the tank, which holds 43 l. How much petrol did I put in?

A: 40·8 l

Q: I have 40·8 l of petrol. I use one quarter. How much petrol have I now?

A: 3·8 l

Q: I have 3·8 l of petrol. I put in 37 l of petrol. How much do I have now?

A: 27·5 l

Q: I have 27·5 l of petrol. I use 3500 ml. How much petrol have I now?

A: 24 l

Q: I have 24 l of petrol. I use three quarters. How much petrol have I now?

A: 30·6 l

Q: I have 30·6 l of petrol. I use one third. How much petrol have I now?

Teachers' note The cards form one continual loop. They can be used for a whole class or group/pair activity, where each child has a card and the questions are read aloud.

Developing Numeracy
Solving Problems Year 5
© A & C Black

Relay runners

Four children in a relay team
run a race.

• **Read the times for each lap.**

Lap 1:	Sally	43·9 seconds
Lap 2:	Bob	51·8 seconds
Lap 3:	Azra	48·7 seconds
Lap 4:	Paul	46·6 seconds

1. What was the total time for all four laps? _____

2. Fill in the children's names in the stopwatches.

Fastest   Next fastest   Next fastest   Next fastest

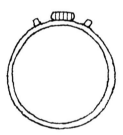 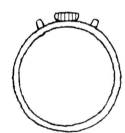 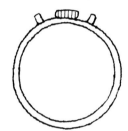 

3. What is the difference between Sally's and Bob's times? _____

4. What is the difference between Azra's and Paul's times? _____

5. How much faster than 50 seconds was Azra? _____

6. How much faster than 50 seconds was Sally? _____

7. David runs around the track in 47·5 seconds. Who might he replace

     in the relay team? _____

Now try this!

• **Work out the average time.** _____

• **How much faster or slower than average was**
     each child?

Teachers' note The children could use calculators to find the answers. Remind them that the
slower times are larger numbers and the faster times are smaller numbers.

**Developing Numeracy
Solving Problems Year 5
© A & C Black**

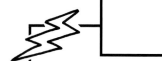

Timetable teaser

Here are the times of four trains from East Grinstead to London.

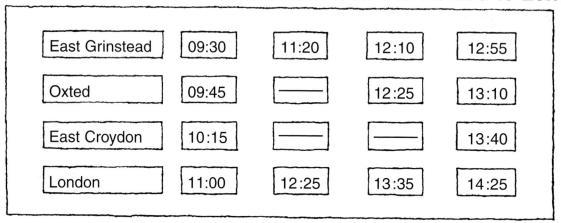

East Grinstead	09:30	11:20	12:10	12:55
Oxted	09:45	——	12:25	13:10
East Croydon	10:15	——	——	13:40
London	11:00	12:25	13:35	14:25

1. At what time does the 09:30 train from East Grinstead arrive in London? _____

2. How long does the 12:55 train from East Grinstead take to reach Oxted? _____

3. How long does the 12:25 train from Oxted take to reach London? _____

4. At how many stations does the 12:10 from East Grinstead stop before it reaches London? _____

5. Which is the fastest train from East Grinstead to London? _____

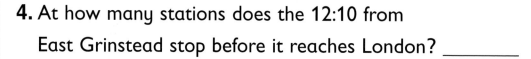

6. If you waited at Oxted at 12:00 for a train to London, what time would you arrive in London? _____

7. You are in Oxted. You have to get to East Croydon by 1 pm. Which train must you catch? _____

8. Which train would you catch from East Grinstead to be in London by 1:30 pm? _____

 • Work out the average journey time from East Grinstead to London.

Teachers' note Begin the lesson with some revision of calculating time differences. Remind the children that a subtraction method is not appropriate, for example 1310 – 1255, as time is in base 60 not base 10.

**Developing Numeracy
Solving Problems Year 5
© A & C Black**

Cake bake

Jenny is making her dad a surprise birthday cake.

I go with Gran to the supermarket to buy the ingredients ($\frac{3}{4}$ hour trip).
I mix up the ingredients ($\frac{1}{2}$ hour).
I bake the cake (45 minutes).
I wait for the cake to cool (35 minutes).

It has to be ready before Dad gets home.

1. How long did Jenny spend on her dad's cake? _____

2. Jenny went to the supermarket at 11:45 am.
What time did she get home? _____

3. Jenny put the cake in the oven at 13:30.
What time did she take it out? _____
By what time had it cooled? _____

4. At 4 pm Jenny puts marzipan and icing on the cake.
Each layer takes 12 minutes. She puts on one layer of marzipan and
two layers of icing. How long does this take? _____
What time does she finish? _____

5. Dad is due to arrive home at 17:45.
How long does Jenny have to wait?

_____

• **If each activity took** 5 minutes **longer, would the cake be ready in time?** _____

Teachers' note Provide the children with small clocks to assist them with this page.

Developing Numeracy
Solving Problems Year 5
© A & C Black

The leisure centre

Judo lessons
8 week course
Monday evenings beginning the
first week of April

Sauna
Opens on the first
Friday in June

Health and Fitness Room

Closed for five weeks
Opens again on 27 May

Rock-climbing Club
We meet each Wednesday at
7.00 pm. Join us!

• **Use a calendar to answer the questions.**

1. What is the date of the first judo lesson?

_____

2. Write the dates of all the rock-climbing meetings in June.

_____

_____

3. On what date did the health and fitness room close?

_____

4. What is the date of the third judo lesson?

_____

5. On what date does the sauna open for the first time?

_____

6. On what date is the last lesson of the judo course?

_____

7. How many rock-climbing meetings are there in September?

_____

8. If the health and fitness room opens 5 days late, what day and date is this?

_____

Teachers' note Revise how the calendar is arranged and provide a calendar of the current year for each child or pair of children. Answers will vary from year to year. For further extension, ask the children to write the days of the week in their answers.

Developing Numeracy
Solving Problems Year 5
© A & C Black

The leisure centre

Squash Club

We meet each fortnight on Wednesday evening.
Starts second week in May.

Beauty Salon

Closed for nine weeks
Opens again on 27 May

Jacuzzi

Opens on the second Friday in June

Swimming lessons

8 week course
Monday evenings beginning the third week of March

• **Use a calendar to answer the questions.**

Write the day **and the** date .

1. When is the first swimming lesson?

_____

2. When does the squash club meet in July and August?

_____

_____

3. When did the beauty salon close?

_____

4. When is the third swimming lesson?

_____

5. When does the jacuzzi open for the first time?

_____

6. When is the last swimming lesson?

_____

7. How many squash club meetings are there in September?

_____

8. If the beauty salon opens 5 days late, when does it open?

_____

Teachers' note Revise how the calendar is arranged and provide a calendar of the current year for each child or pair of children. Answers will vary from year to year. The children could create their own questions or statements about the dates of the leisure centre's activities.

Developing Numeracy
Solving Problems Year 5
© A & C Black

p 6
1. x
2. ÷
3. −
4. +
5. ÷
6. ÷
7. x
8. +
9. +
10. −
11. x
12. −

p 7
1. 98 × 64 = 6272
2. 80 × 30 = 2400
3. 61 × 47 = 2867
4. 19 × 84 = 1596
5. 46 × 13 = 598
6. 24 × 94 = 2256
7. 81 × 42 =3402
8. 23 × 65 = 1495

p 8
1. 325
2. 8
3. 180
4. 210
5. 3 r 11
6. 4028
7. 49·5

p 10
There is a range of methods. For reference, the following use each number only once.
1. (100 + 10) × 8 + 1
2. 3 × 100 + 8 − 5
3. 100 + (10 × 5) + 3
4. 100 + (10 × 5) − 3
5. (100 + 10) × 5 + 8 − 1
6. 10 × (5 + 3) + (50 × 2) + 1
7. (50 × 5) + 10 + 3 + 2 + 1
8. (3 + 5) × 10 + 50 + 2
9. (50 + 10) × 2
10. 50 × (5 + 2) + 1

p 11
There is a range of methods which include:
1. (8 × 2 × 50) − 5 − 1
2. (50 × 10) + (5 × 8) + 2
3. (50 × 8) + 5 + 2 + 1
4. (50 + 10 + 2) × (8 − 5)
5. (50 × 2) × 8 + (5 × 10) + 1
6. (75 × 10) + 1 + 6
7. (75 × 4) + (6 × 10) + 1
8. [(6 + 4) × 10] + 9
9. (75 × 4) + (9 × 10) + 6 + 1
10. (75 × 6) + 9 + 1 − 4

p 12

8	1	2
4	3	7
9	6	5

p 13
1.

3	8	5
2	9	1
6	4	7

2.

1	2	9
3	6	7
4	8	5

3.

7	2	5
1	4	8
3	6	9

p 14
1. £2.54
2. A bit each day, doubling
3. One amount
4. A bit each day, doubling

Now try this!
A bit each day, doubling

p 15
1. 9, 16, 21, 24, 25
2. 8, 14, 18, 20, 24, 30, 32, 36

Now try this!
7, 12, 15, 16, 20, 24, 27, 32, 36

p 16
1. 93, 94
2. 17, 18
3. 71, 72, 73
4. 35, 36, 37
5. 39, 40, 41, 42
6. 98, 99, 100, 101
7. 11, 12
8. 13, 14
9. 24, 25
10. 9, 10, 11

p 17
Now try this!
1, 1, 6, 6
1, 1, 8, 4
1, 1, 10, 2 etc.

p 18
There are various solutions.

Now try this!
120

p 19
There are various solutions. Note that 167 with 3 darts is not possible.

Now try this!
180

p 20
Pattern for 3:   Pattern for 5:

Now try this!
Pattern for 8:

p 21
1. True
2. False
3. True
4. True

p 24
1. 2, 5, 8, 11, 14, 17, 20, 23, 26, 29, 32
2. 128, 124, 120, 116, 112, 108, 104, 100, 96, 92, 88
3. 26, 21, 16, 11, 6, 1, −4, −9, −14, −19, −24
4. 1, 4, 9, 16, 25, 36, 49, 64, 81, 100, 121

Now try this!
1, 3, 6, 10, 15, 21, 28… These are triangular numbers.

p 25
1. 2, 4, 6, 8, 10, 12, 14, 16, 18, 20, 22
   20th number = 40
2. 100, 95, 90, 85, 80, 75, 70, 65, 60, 55, 50
   20th number = 5
3. 1, 4, 9, 16, 25, 36, 49, 64, 81, 100, 121
   20th number = 400

Now try this!
3, 6, 9, 12, 15, 18…
10th number = 30

p 26
1. 13   19   31
2. 11   15   23

Now try this!
18   30   300

p 27
Now try this!

Number of peaches	1	2	3	4	5
Number of arrangements	6	15	19	15	6

There are the same number of arrangements for 1 and 5 peaches, and for 2 and 4 peaches.

p 28
1. Brown bread and jam
   Brown bread and chocolate spread
   Brown bread and cheese
   White bread and jam
   White bread and chocolate spread
   White bread and cheese
2. 6
3. 3
4. 9

Now try this!
There are nine combinations:
   Brown bread, jam and cheese
   Brown bread, jam and chocolate spread
   Brown bread, cheese and chocolate spread
   White bread, jam and cheese
   White bread, jam and chocolate spread
   White bread, cheese and chocolate spread
   Wholemeal bread, jam and cheese
   Wholemeal bread, jam and chocolate spread
   Wholemeal bread, cheese and chocolate spread

p 29
1. 12
2. 12
3. 15

Now try this!
14 squares and 22 rectangles

p 30
1. and 2.
There are many possibilities. Ensure that the children have not included mirror images.

Now try this!
There are many possibilities, e.g.

 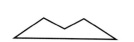

p 31
1. and 2.
There are many possibilities. Ensure that the children have not included mirror images.

Now try this!
There are many possibilities, e.g.

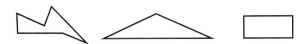

p 32
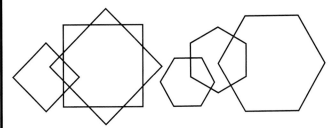

p 33
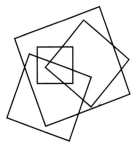 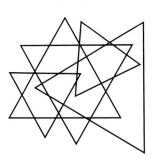

p 34
Yes, every pair of angles on a straight line add up to 180 degrees.

p 35
The perimeter of any regular shape can be found by multiplying the number of sides by the length of one side.

p 37
1. 376
2. 121
3. 3
4. 409
5. about 7
6. about 19
7. 272
8. 97
9. 3
10. 106

Now try this!
About 7

p 39
1. £6, £15
2. £2, £5
3. £23
4. £13.01
5. £4.80, £12, £31.20, £62.40

Now try this!
£300, £1800, £18 000

p 40
1. 10p, 30p
2. 15p, 30p
3. 30p, 35p
4. 31p, 26p

p 41
1. 25p, 60p
2. 40p, 60p
3. 10p, 80p
4. 45p, 5p, 90p

p 42
1. £640
2. £18
3. £60
4. £200
5. £3, £4.50, £7.50, £30

p 43
1. A = £24   B = £23
   C = £21   D = £18
   cheapest = D; most expensive = A
2. A = £11   B = £63
   C = £12   D = £18
   cheapest = A; most expensive = B
3. A = £8.50   B = £10.50
   C = £4.50   D = £20
   cheapest = C; most expensive = D

Now try this!
£5, £1.20, 30p (or £0.30), 6p (or £0.06)

p 44
1. 10 dollars, 32 dollars, 76 dollars, 190 dollars
2. 45 francs, 63 francs, 99 francs, 1080 francs
3. 1500 drachmas, 4000 drachmas, 6000 drachmas, 30 000 drachmas

Now try this!
100 dollars, 450 francs, 25 000 drachmas

p 45
1. 3.2 dollars, 6.4 dollars, 32 dollars, 128 dollars
2. 17.4 francs, 43.5 francs, 217.5 francs, 1044 francs
3. 1350 drachmas, 3600 drachmas, 5400 drachmas, 40 500 drachmas

Now try this!
80 dollars, 435 francs, 22 500 drachmas

p 46
1. 2 days = 5 hours
1 week = 17½ hours
2 weeks = 35 hours
1 month = 70 hours (4 weeks) or 75 hours (30 days)
or 77½ hours (31 days)
1 year = 912½ hours or an approximation
2. 08:25
3. 19:10
4. 1 hour 15 minutes
5. 2 minutes
6. 54 minutes
Now try this!
3·75 km or 3750 m

p 47
1. 2 days = £2.40
1 week = £8.40
2 weeks = £16.80
1 month = £33.60 (4 weeks) or £36 (30 days) or £37.20 (31 days)
1 year = £438 or an approximation
2. 18/19 June
3. 11 weeks
4. 77 days
5. 192½ hours
Now try this!
5 times

p 48
1. Yes
2. 40 cm
3. No
4. 1·56 m or 156 cm
5. 1·44 m or 144 cm
6. 97 cm
7. 6 cm
8. 128 cm
Now try this!
42 presents

p 49
1. 524 miles
2. 384 miles
3. 129 miles
4. 1·4 km
5. 8·1 km
6. 900 m

p 50
1. 11 m
2. 50 cm or 0·5 m
3. 160 cm or 1·6 m
4. 285 cm, 2·85 m
5. 65 cm, 0·65 m
6. 170 cm or 1·7 m
Now try this!
136 cm, 136 cm, 52 cm

p 51
1. 1550 g or 1·55 kg
2. 38·3 kg
3. 55 g
4. 300 g
5. 13
6. 480 kg
7. 6·5 kg
8. 22 kg
9. 325 g
10. 875 g

p 52
Shortbread
Total grams = 520 g
60 g
45 g
25 g

Gingernuts
Total grams = 260 g
25 g
10 g
15 g
2 g

Pasta sauce
Total grams = 615 g
Approx. 63 g
Approx. 22 g
Approx. 12/13 g
5 g
Now try this!
270 g flour
150 g fat
120 g sugar
180 g chocolate

p 53
1. 1600 g
2. 250 g
3. 75 g
4. 420 g, 1 tomato
5. 4, 20, 40
6. 6
Now try this!
500 g, 200 g, 143 g, 83 g, 71 g

p 54
1. 902 ml
2. 98 ml
3. 45
4. 180
5. 451 ml
6. 28, 56
280, 21
Now try this!
250 ml, 500 ml

p 55
1. about 14
2. 33
3. 570 ml
4. 30
5. about 263 ml
6. about 16 ml
Now try this!
45 litres, 45·09 litres

p 57
1. 3 minutes 11 seconds (191 seconds)
2. Sally, Paul, Azra, Bob
3. 7·9 seconds
4. 2·1 seconds
5. 1·3 seconds
6. 6·1 seconds
7. Bob or Azra
Now try this!
Average time = 47·75 seconds
Sally was 3·85 seconds faster.
Bob was 4·05 seconds slower.
Azra was 0·95 seconds slower.
Paul was 1·15 seconds faster.

p 58
1. 11:00
2. 15 minutes
3. 1 hour 10 minutes
4. 1
5. 11:20
6. 13:35
7. 09:45
8. 11:20
Now try this!
1 hour 22½ minutes

p 59
1. 2 hours 35 minutes
2. 12:30
3. 14:15, 14:50
4. 36 minutes, 4:36 pm
5. 1 hour 9 minutes
Now try this!
Yes

p 60 and p 61
Dates will vary depending on year.